TECH PROPHECIES
2100 TO 2199

I0407351

Year=1992, Start 81 year cycle for A1 diagram

Year+54=1992, Start 81 year cycle for A2 diagram

Year+27=1992, Start 81 year cycle for B1 diagram

Year=1992, Start 81 year cycle for B2 diagram

Year+54=1992, Start 81 year cycle for C1 diagram

Year+27=1992, Start 81 year cycle for C2 diagram

TECH PROPHECIES 2100 TO 2199 / NATHAN COPPEDGE

, , ,

An 'excel' version is available @ :

https://drive.google.com/file/d/13yR_Bo5gEmFlr
yedTD0wWb8EsVmV-TXZ/view

TECH PROPHECIES
2100 TO 2199

ACCURATE TECHNOLOGY PREDICTIONS FOR THE 22ND CENTURY

NATHAN COPPEDGE

TECH PROPHECIES 2100 TO 2199 / NATHAN COPPEDGE

,,,

INTRO

These are tech predictions beginning in 2100 and continuing to 2199. They are divided into 81-year epochs I call 'Copernican Ages of Technology'.

It may be predicted that this part of history includes the concluding events of the Age of Immortality, Perpetual Motion, and Immunity which was the previous Copernican Age ending in 2130, with the new age beyond immortality noted in 2131.

These are tech and discovery predictions for every type of innovation in any species, and even may be used to help prediction natural events.

COVERING YEARS:

2100, 2101, 2102, 2103, 2104, 2105, 2106, 2107, 2108, 2109, 2110, 2111, 2112, 2113, 2114, 2115, 2116, 2117, 2118, 2119, 2120, 2121, 2122, 2123, 2124, 2125, 2126, 2127, 2128, 2129, 2130, 2131, 2132, 2133, 2134, 2135, 2136, 2137, 2138, 2139, 2140, 2141, 2142, 2143, 2144, 2145, 2146, 2147, 2148, 2149, 2150, 2151, 2152, 2153, 2154, 2155, 2156, 2157, 2158, 2159, 2160, 2161, 2162, 2163, 2164, 2165, 2166, 2167, 2168, 2169, 2170, 2171, 2172, 2173, 2174, 2175, 2176, 2177, 2178, 2179, 2180, 2181, 2182, 2183, 2184, 2185, 2186, 2187, 2188, 2189, 2190, 2191, 2192, 2193, 2194, 2195, 2196, 2197, 2198, 2199

TECH PROPHECIES 2100 TO 2199 / NATHAN COPPEDGE

,,,

DISCOVERY PREDICTIONS

YEAR =   2100 INPUT

A1: COSM SIMPLE COSM SIMPLE

A2: SUBJECTIVE COMPLEX MATH COMPLEX

B1: INTELLIGENT COMPLEXITY UNIVERSAL TECH

B2: SPECIAL BASIC SPECIAL BASIC

C1: UNIVERSAL UNIVERSE CREATIVE UNIVERSE

C2: BASIC TECHNIQUE SUBJECTIVE SUBJECTIVE

YEAR =   2101 INPUT

A1: PHYS SIMPLE COSM SIMPLE

A2: ABSTRACT COMPLEX MATH COMPLEX

B1: INTELLIGENT COMPLEXITY PHYS TECH

B2: ABSTRACT BASIC SPECIAL BASIC

C1: PHYS UNIVERSE CREATIVE UNIVERSE

C2: BASIC TECHNIQUE ABSTRACT SUBJECTIVE

YEAR =   2102 INPUT

A1: PHYS COMPLEX COSM SIMPLE

A2: ABSTRACT SIMPLE MATH COMPLEX

B1: INTELLIGENT COMPLEXITY PHYS CREATIVE

B2: ABSTRACT MATH SPECIAL BASIC

C1: PHYS PHYS CREATIVE UNIVERSE

C2: BASIC TECHNIQUE ABSTRACT ABSTRACT

TECH PROPHECIES 2100 TO 2199 / NATHAN COPPEDGE

YEAR =   2103 INPUT
- A1: NEW INVENTION COSM SIMPLE
- A2: GROUP CONCEPT MATH COMPLEX
- B1: INTELLIGENT COMPLEXITY INTELLIGENT COMPLEXITY END OF A COPERNICAN AGE
- B2: STUPIDLY SIMPLE SPECIAL BASIC
- C1: INTELLIGENT CREATIVE CREATIVE UNIVERSE
- C2: BASIC TECHNIQUE BASIC TECHNIQUE END OF A COPERNICAN AGE

YEAR =   2104 INPUT
- A1: PHYS SIMPLE TECH COMPLEX
- A2: MATH SIMPLE ARCHAIC SIMPLE
- B1: TECH CREATIVE TECH CREATIVE BEGINNING OF A NEW COPERNICAN AGE
- B2: ABSTRACT BASIC BASIC MATH
- C1: CREATIVE PHYS TECH PHYS
- C2: SIMPLE ABSTRACT SIMPLE ABSTRACT BEGINNING OF A NEW COPERNICAN AGE

YEAR =   2105 INPUT
- A1: PHYS SIMPLE TECH SIMPLE
- A2: MATH SIMPLE ARCHAIC COMPLEX
- B1: TECH TECH TECH CREATIVE
- B2: ABSTRACT BASIC BASIC BASIC
- C1: CREATIVE PHYS TECH UNIVERSE
- C2: SIMPLE SUBJECTIVE SIMPLE ABSTRACT

YEAR =   2106 INPUT
   A1: PHYS SIMPLE ART SIMPLE
   A2: MATH SIMPLE MATH COMPLEX
   B1: CREATIVE TECH TECH CREATIVE
   B2: ABSTRACT BASIC MATH BASIC
   C1: CREATIVE PHYS CREATIVE UNIVERSE
   C2: MATH SUBJECTIVE SIMPLE ABSTRACT

YEAR =   2107 INPUT
   A1: PHYS SIMPLE ART COMPLEX
   A2: MATH SIMPLE MATH SIMPLE
   B1: CREATIVE CREATIVE TECH CREATIVE
   B2: ABSTRACT BASIC MATH MATH
   C1: CREATIVE PHYS CREATIVE PHYS
   C2: MATH ABSTRACT SIMPLE ABSTRACT

YEAR =   2108 INPUT
   A1: PHYS SIMPLE COSM COMPLEX
   A2: SUBJECTIVE SIMPLE MATH SIMPLE
   B1: UNIVERSAL CREATIVE TECH CREATIVE
   B2: ABSTRACT BASIC SPECIAL MATH
   C1: UNIVERSAL PHYS CREATIVE PHYS
   C2: SUBJECTIVE ABSTRACT SIMPLE ABSTRACT

YEAR =   2109 INPUT
A1: PHYS SIMPLE COSM SIMPLE
A2: SUBJECTIVE COMPLEX MATH SIMPLE
B1: UNIVERSAL TECH TECH CREATIVE
B2: ABSTRACT BASIC SPECIAL BASIC
C1: UNIVERSAL UNIVERSE CREATIVE PHYS
C2: SUBJECTIVE SUBJECTIVE SIMPLE ABSTRACT

YEAR =   2110 INPUT
A1: PHYS SIMPLE PHYS SIMPLE
A2: ABSTRACT COMPLEX MATH SIMPLE
B1: PHYS TECH TECH CREATIVE
B2: ABSTRACT BASIC ABSTRACT BASIC
C1: PHYS UNIVERSE CREATIVE PHYS
C2: ABSTRACT SUBJECTIVE SIMPLE ABSTRACT

YEAR =   2111 INPUT
A1: PHYS COMPLEX PHYS SIMPLE
A2: ABSTRACT SIMPLE MATH SIMPLE
B1: PHYS CREATIVE TECH CREATIVE
B2: ABSTRACT MATH ABSTRACT BASIC
C1: PHYS PHYS CREATIVE PHYS
C2: ABSTRACT ABSTRACT SIMPLE ABSTRACT

YEAR =   2112 INPUT
- A1: NEW INVENTION PHYS SIMPLE
- A2: GROUP CONCEPT MATH SIMPLE
- B1: INTELLIGENT COMPLEXITY TECH CREATIVE
- B2: STUPIDLY SIMPLE ABSTRACT BASIC
- C1: INTELLIGENT CREATIVE CREATIVE PHYS
- C2: BASIC TECHNIQUE SIMPLE ABSTRACT

YEAR =   2113 INPUT
- A1: PHYS COMPLEX TECH COMPLEX
- A2: SUBJECTIVE SIMPLE ARCHAIC SIMPLE
- B1: TECH TECH TECH CREATIVE
- B2: ABSTRACT MATH BASIC MATH
- C1: UNIVERSAL PHYS TECH PHYS
- C2: SIMPLE SUBJECTIVE SIMPLE ABSTRACT

YEAR =   2114 INPUT
- A1: PHYS COMPLEX TECH SIMPLE
- A2: SUBJECTIVE SIMPLE ARCHAIC COMPLEX
- B1: TECH TECH TECH TECH
- B2: ABSTRACT MATH BASIC BASIC
- C1: UNIVERSAL PHYS TECH UNIVERSE
- C2: SIMPLE SUBJECTIVE SIMPLE SUBJECTIVE

YEAR =   2115 INPUT
- A1: **PHYS COMPLEX ART SIMPLE**
- A2: **SUBJECTIVE SIMPLE MATH COMPLEX**
- B1: **CREATIVE TECH TECH TECH**
- B2: **ABSTRACT MATH MATH BASIC**
- C1: **UNIVERSAL PHYS CREATIVE UNIVERSE**
- C2: **MATH SUBJECTIVE SIMPLE SUBJECTIVE**

YEAR =   2116 INPUT
- A1: **PHYS COMPLEX ART COMPLEX**
- A2: **SUBJECTIVE SIMPLE MATH SIMPLE**
- B1: **CREATIVE TECH TECH CREATIVE**
- B2: **ABSTRACT MATH MATH MATH**
- C1: **UNIVERSAL PHYS CREATIVE PHYS**
- C2: **MATH SUBJECTIVE SIMPLE ABSTRACT**

YEAR =   2117 INPUT
- A1: **PHYS COMPLEX COSM COMPLEX**
- A2: **SUBJECTIVE SIMPLE SUBJECTIVE SIMPLE**
- B1: **UNIVERSAL CREATIVE TECH TECH**
- B2: **ABSTRACT MATH SPECIAL MATH**
- C1: **UNIVERSAL PHYS UNIVERSAL PHYS**
- C2: **SUBJECTIVE ABSTRACT SIMPLE SUBJECTIVE**

YEAR =   2118 INPUT
- A1: PHYS COMPLEX COSM SIMPLE
- A2: SUBJECTIVE COMPLEX SUBJECTIVE SIMPLE
- B1: UNIVERSAL TECH TECH TECH
- B2: ABSTRACT MATH SPECIAL BASIC
- C1: UNIVERSAL UNIVERSE UNIVERSAL PHYS
- C2: SUBJECTIVE SUBJECTIVE SIMPLE SUBJECTIVE

YEAR =   2119 INPUT
- A1: PHYS COMPLEX PHYS SIMPLE
- A2: ABSTRACT COMPLEX SUBJECTIVE SIMPLE
- B1: PHYS TECH TECH TECH
- B2: ABSTRACT MATH ABSTRACT BASIC
- C1: PHYS UNIVERSE UNIVERSAL PHYS
- C2: ABSTRACT SUBJECTIVE SIMPLE SUBJECTIVE

YEAR =   2120 INPUT
- A1: PHYS COMPLEX PHYS COMPLEX
- A2: ABSTRACT SIMPLE SUBJECTIVE SIMPLE
- B1: PHYS CREATIVE TECH TECH
- B2: ABSTRACT MATH ABSTRACT MATH
- C1: PHYS PHYS UNIVERSAL PHYS
- C2: ABSTRACT ABSTRACT SIMPLE SUBJECTIVE

YEAR =   2121 INPUT
    A1: NEW INVENTION PHYS COMPLEX
    A2: GROUP CONCEPT SUBJECTIVE SIMPLE
    B1: INTELLIGENT COMPLEXITY TECH TECH
    B2: STUPIDLY SIMPLE ABSTRACT MATH
    C1: INTELLIGENT CREATIVE UNIVERSAL PHYS
    C2: BASIC TECHNIQUE SIMPLE SUBJECTIVE

YEAR =   2122 INPUT
    A1: NEW INVENTION TECH COMPLEX
    A2: SUBJECTIVE COMPLEX ARCHAIC SIMPLE
    B1: CREATIVE TECH TECH CREATIVE
    B2: STUPIDLY SIMPLE BASIC MATH
    C1: UNIVERSAL UNIVERSE TECH PHYS
    C2: MATH SUBJECTIVE SIMPLE ABSTRACT

YEAR =   2123 INPUT
    A1: NEW INVENTION TECH SIMPLE
    A2: SUBJECTIVE COMPLEX ARCHAIC COMPLEX
    B1: CREATIVE TECH TECH TECH
    B2: STUPIDLY SIMPLE BASIC BASIC
    C1: UNIVERSAL UNIVERSE TECH UNIVERSE
    C2: MATH SUBJECTIVE SIMPLE SUBJECTIVE

YEAR =   2124 INPUT
    A1: NEW INVENTION ART SIMPLE
    A2: SUBJECTIVE COMPLEX MATH COMPLEX
    B1: CREATIVE TECH CREATIVE TECH
    B2: STUPIDLY SIMPLE MATH BASIC
    C1: UNIVERSAL UNIVERSE CREATIVE UNIVERSE
    C2: MATH SUBJECTIVE MATH SUBJECTIVE

YEAR =   2125 INPUT
    A1: NEW INVENTION ART COMPLEX
    A2: SUBJECTIVE COMPLEX MATH SIMPLE
    B1: CREATIVE CREATIVE CREATIVE TECH
    B2: STUPIDLY SIMPLE MATH MATH
    C1: UNIVERSAL UNIVERSE CREATIVE PHYS
    C2: MATH ABSTRACT MATH SUBJECTIVE

YEAR =   2126 INPUT
    A1: NEW INVENTION COSM COMPLEX
    A2: SUBJECTIVE COMPLEX SUBJECTIVE SIMPLE
    B1: UNIVERSAL CREATIVE CREATIVE TECH
    B2: STUPIDLY SIMPLE SPECIAL MATH
    C1: UNIVERSAL UNIVERSE UNIVERSAL PHYS
    C2: SUBJECTIVE ABSTRACT MATH SUBJECTIVE

TECH PROPHECIES 2100 TO 2199 / NATHAN COPPEDGE

YEAR =   2127 INPUT
    A1: NEW INVENTION COSM SIMPLE
    A2: SUBJECTIVE COMPLEX SUBJECTIVE COMPLEX
    B1: UNIVERSAL TECH CREATIVE TECH
    B2: STUPIDLY SIMPLE SPECIAL BASIC
    C1: UNIVERSAL UNIVERSE UNIVERSAL UNIVERSE
    C2: SUBJECTIVE SUBJECTIVE MATH SUBJECTIVE

YEAR =   2128 INPUT
    A1: NEW INVENTION PHYS SIMPLE
    A2: ABSTRACT COMPLEX SUBJECTIVE COMPLEX
    B1: PHYS TECH CREATIVE TECH
    B2: STUPIDLY SIMPLE ABSTRACT BASIC
    C1: PHYS UNIVERSE UNIVERSAL UNIVERSE
    C2: ABSTRACT SUBJECTIVE MATH SUBJECTIVE

YEAR =   2129 INPUT
    A1: NEW INVENTION PHYS COMPLEX
    A2: ABSTRACT SIMPLE SUBJECTIVE COMPLEX
    B1: PHYS CREATIVE CREATIVE TECH
    B2: STUPIDLY SIMPLE ABSTRACT MATH
    C1: PHYS PHYS UNIVERSAL UNIVERSE
    C2: ABSTRACT ABSTRACT MATH SUBJECTIVE

YEAR =   2130 INPUT
  A1: NEW INVENTION NEW INVENTION END OF A COPERNICAN AGE
  A2: GROUP CONCEPT SUBJECTIVE COMPLEX
  B1: INTELLIGENT COMPLEXITY CREATIVE TECH
  B2: STUPIDLY SIMPLE STUPIDLY SIMPLE END OF COPERNICAN AGE
  C1: INTELLIGENT CREATIVE UNIVERSAL UNIVERSE
  C2: BASIC TECHNIQUE MATH SUBJECTIVE

YEAR =   2131 INPUT
  A1: TECH COMPLEX TECH COMPLEX BEGINNING OF A NEW COPERNICAN AGE
  A2: ABSTRACT COMPLEX ARCHAIC SIMPLE
  B1: CREATIVE CREATIVE TECH CREATIVE
  B2: BASIC MATH BASIC MATH BEGINNING OF A NEW COPERNICAN AGE
  C1: PHYS UNIVERSE TECH PHYS
  C2: MATH ABSTRACT SIMPLE ABSTRACT

YEAR =   2132 INPUT
  A1: TECH SIMPLE TECH COMPLEX
  A2: ABSTRACT COMPLEX ARCHAIC COMPLEX
  B1: CREATIVE CREATIVE TECH TECH
  B2: BASIC BASIC BASIC MATH
  C1: PHYS UNIVERSE TECH UNIVERSE
  C2: MATH ABSTRACT SIMPLE SUBJECTIVE

YEAR =   2133 INPUT
    A1: ART SIMPLE TECH COMPLEX
    A2: ABSTRACT COMPLEX MATH COMPLEX
    B1: CREATIVE CREATIVE CREATIVE TECH
    B2: MATH BASIC BASIC MATH
    C1: PHYS UNIVERSE CREATIVE UNIVERSE
    C2: MATH ABSTRACT MATH SUBJECTIVE

YEAR =   2134 INPUT
    A1: ART COMPLEX TECH COMPLEX
    A2: ABSTRACT COMPLEX MATH SIMPLE
    B1: CREATIVE CREATIVE CREATIVE CREATIVE
    B2: MATH MATH BASIC MATH
    C1: PHYS UNIVERSE CREATIVE PHYS
    C2: MATH ABSTRACT MATH ABSTRACT

YEAR =   2135 INPUT
    A1: COSM COMPLEX TECH COMPLEX
    A2: ABSTRACT COMPLEX SUBJECTIVE SIMPLE
    B1: UNIVERSAL CREATIVE CREATIVE CREATIVE
    B2: SPECIAL MATH BASIC MATH
    C1: PHYS UNIVERSE UNIVERSAL PHYS
    C2: SUBJECTIVE ABSTRACT MATH ABSTRACT

YEAR =   2136 INPUT
- A1: COSM SIMPLE TECH COMPLEX
- A2: ABSTRACT COMPLEX SUBJECTIVE COMPLEX
- B1: UNIVERSAL TECH CREATIVE CREATIVE
- B2: SPECIAL BASIC BASIC MATH
- C1: PHYS UNIVERSE UNIVERSAL UNIVERSE
- C2: SUBJECTIVE SUBJECTIVE MATH ABSTRACT

YEAR =   2137 INPUT
- A1: PHYS SIMPLE TECH COMPLEX
- A2: ABSTRACT COMPLEX ABSTRACT COMPLEX
- B1: PHYS TECH CREATIVE CREATIVE
- B2: ABSTRACT BASIC BASIC MATH
- C1: PHYS UNIVERSE PHYS UNIVERSE
- C2: ABSTRACT SUBJECTIVE MATH ABSTRACT

YEAR =   2138 INPUT
- A1: PHYS COMPLEX TECH COMPLEX
- A2: ABSTRACT SIMPLE ABSTRACT COMPLEX
- B1: PHYS CREATIVE CREATIVE CREATIVE
- B2: ABSTRACT MATH BASIC MATH
- C1: PHYS PHYS PHYS UNIVERSE
- C2: ABSTRACT ABSTRACT MATH ABSTRACT

YEAR =   2139 INPUT
   A1: NEW INVENTION TECH COMPLEX
   A2: GROUP CONCEPT ABSTRACT COMPLEX
   B1: INTELLIGENT COMPLEXITY CREATIVE CREATIVE
   B2: STUPIDLY SIMPLE BASIC MATH
   C1: INTELLIGENT CREATIVE PHYS UNIVERSE
   C2: BASIC TECHNIQUE MATH ABSTRACT

YEAR =   2140 INPUT
   A1: TECH SIMPLE TECH COMPLEX
   A2: ABSTRACT SIMPLE ARCHAIC SIMPLE
   B1: UNIVERSAL CREATIVE TECH CREATIVE
   B2: BASIC BASIC BASIC MATH
   C1: PHYS PHYS TECH PHYS
   C2: SUBJECTIVE ABSTRACT SIMPLE ABSTRACT

YEAR =   2141 INPUT
   A1: TECH SIMPLE TECH SIMPLE
   A2: ABSTRACT SIMPLE ARCHAIC COMPLEX
   B1: UNIVERSAL CREATIVE TECH TECH
   B2: BASIC BASIC BASIC BASIC
   C1: PHYS PHYS TECH UNIVERSE
   C2: SUBJECTIVE ABSTRACT SIMPLE SUBJECTIVE

YEAR =   2142 INPUT
    A1: ART SIMPLE TECH SIMPLE
    A2: ABSTRACT SIMPLE MATH COMPLEX
    B1: UNIVERSAL CREATIVE CREATIVE TECH
    B2: MATH BASIC BASIC BASIC
    C1: PHYS PHYS CREATIVE UNIVERSE
    C2: SUBJECTIVE ABSTRACT MATH SUBJECTIVE

YEAR =   2143 INPUT
    A1: ART SIMPLE TECH COMPLEX
    A2: ABSTRACT SIMPLE MATH SIMPLE
    B1: UNIVERSAL CREATIVE CREATIVE CREATIVE
    B2: MATH BASIC BASIC MATH
    C1: PHYS PHYS CREATIVE PHYS
    C2: SUBJECTIVE ABSTRACT MATH ABSTRACT

YEAR =   2144 INPUT
    A1: COSM COMPLEX TECH SIMPLE
    A2: ABSTRACT SIMPLE SUBJECTIVE SIMPLE
    B1: UNIVERSAL CREATIVE UNIVERSAL CREATIVE
    B2: SPECIAL MATH BASIC BASIC
    C1: PHYS PHYS UNIVERSAL PHYS
    C2: SUBJECTIVE ABSTRACT SUBJECTIVE ABSTRACT

YEAR =   2145 INPUT
   A1: COSM SIMPLE TECH SIMPLE
   A2: ABSTRACT SIMPLE SUBJECTIVE COMPLEX
   B1: UNIVERSAL TECH UNIVERSAL CREATIVE
   B2: SPECIAL BASIC BASIC BASIC
   C1: PHYS PHYS UNIVERSAL UNIVERSE
   C2: SUBJECTIVE SUBJECTIVE SUBJECTIVE ABSTRACT

YEAR =   2146 INPUT
   A1: PHYS SIMPLE TECH SIMPLE
   A2: ABSTRACT SIMPLE ABSTRACT COMPLEX
   B1: PHYS TECH UNIVERSAL CREATIVE
   B2: ABSTRACT BASIC BASIC BASIC
   C1: PHYS PHYS PHYS UNIVERSE
   C2: ABSTRACT SUBJECTIVE SUBJECTIVE ABSTRACT

YEAR =   2147 INPUT
   A1: PHYS COMPLEX TECH SIMPLE
   A2: ABSTRACT SIMPLE ABSTRACT SIMPLE
   B1: PHYS CREATIVE UNIVERSAL CREATIVE
   B2: ABSTRACT MATH BASIC BASIC
   C1: PHYS PHYS PHYS PHYS
   C2: ABSTRACT ABSTRACT SUBJECTIVE ABSTRACT

YEAR =   2148 INPUT
  A1: NEW INVENTION TECH SIMPLE
  A2: GROUP CONCEPT ABSTRACT SIMPLE
  B1: INTELLIGENT COMPLEXITY UNIVERSAL CREATIVE
  B2: STUPIDLY SIMPLE BASIC BASIC
  C1: INTELLIGENT CREATIVE PHYS PHYS
  C2: BASIC TECHNIQUE SUBJECTIVE ABSTRACT

YEAR =   2149 INPUT
  A1: ART SIMPLE TECH COMPLEX
  A2: GROUP CONCEPT ARCHAIC SIMPLE
  B1: UNIVERSAL TECH TECH CREATIVE
  B2: MATH BASIC BASIC MATH
  C1: INTELLIGENT CREATIVE TECH PHYS
  C2: SUBJECTIVE SUBJECTIVE SIMPLE ABSTRACT

YEAR =   2150 INPUT
  A1: ART SIMPLE TECH SIMPLE
  A2: GROUP CONCEPT ARCHAIC COMPLEX
  B1: UNIVERSAL TECH TECH TECH
  B2: MATH BASIC BASIC BASIC
  C1: INTELLIGENT CREATIVE TECH UNIVERSE
  C2: SUBJECTIVE SUBJECTIVE SIMPLE SUBJECTIVE

YEAR =   2151 INPUT
- A1: ART SIMPLE ART SIMPLE
- A2: GROUP CONCEPT MATH COMPLEX
- B1: UNIVERSAL TECH CREATIVE TECH
- B2: MATH BASIC MATH BASIC
- C1: INTELLIGENT CREATIVE CREATIVE UNIVERSE
- C2: SUBJECTIVE SUBJECTIVE MATH SUBJECTIVE

YEAR =   2152 INPUT
- A1: ART COMPLEX ART SIMPLE
- A2: GROUP CONCEPT MATH SIMPLE
- B1: UNIVERSAL TECH CREATIVE CREATIVE
- B2: MATH MATH MATH BASIC
- C1: INTELLIGENT CREATIVE CREATIVE PHYS
- C2: SUBJECTIVE SUBJECTIVE MATH ABSTRACT

YEAR =   2153 INPUT
- A1: COSM COMPLEX ART SIMPLE
- A2: GROUP CONCEPT SUBJECTIVE SIMPLE
- B1: UNIVERSAL TECH UNIVERSAL CREATIVE
- B2: SPECIAL MATH MATH BASIC
- C1: INTELLIGENT CREATIVE UNIVERSAL PHYS
- C2: SUBJECTIVE SUBJECTIVE SUBJECTIVE ABSTRACT

YEAR =   2154 INPUT
   A1: COSM SIMPLE ART SIMPLE
   A2: GROUP CONCEPT SUBJECTIVE COMPLEX
   B1: UNIVERSAL TECH UNIVERSAL TECH
   B2: SPECIAL BASIC MATH BASIC
   C1: INTELLIGENT CREATIVE UNIVERSAL UNIVERSE
   C2: SUBJECTIVE SUBJECTIVE SUBJECTIVE SUBJECTIVE

YEAR =   2155 INPUT
   A1: PHYS SIMPLE ART SIMPLE
   A2: GROUP CONCEPT ABSTRACT COMPLEX
   B1: PHYS TECH UNIVERSAL TECH
   B2: ABSTRACT BASIC MATH BASIC
   C1: INTELLIGENT CREATIVE PHYS UNIVERSE
   C2: ABSTRACT SUBJECTIVE SUBJECTIVE SUBJECTIVE

YEAR =   2156 INPUT
   A1: PHYS COMPLEX ART SIMPLE
   A2: GROUP CONCEPT ABSTRACT SIMPLE
   B1: PHYS CREATIVE UNIVERSAL TECH
   B2: ABSTRACT MATH MATH BASIC
   C1: INTELLIGENT CREATIVE PHYS PHYS
   C2: ABSTRACT ABSTRACT SUBJECTIVE SUBJECTIVE

TECH PROPHECIES 2100 TO 2199 / NATHAN COPPEDGE

YEAR =   2157 INPUT
- A1: NEW INVENTION ART SIMPLE
- A2: GROUP CONCEPT GROUP CONCEPT END OF A COPERNICAN AGE
- B1: INTELLIGENT COMPLEXITY UNIVERSAL TECH
- B2: STUPIDLY SIMPLE MATH BASIC
- C1: INTELLIGENT CREATIVE INTELLIGENT CREATIVE END OF A COPERNICAN AGE
- C2: BASIC TECHNIQUE SUBJECTIVE SUBJECTIVE

YEAR =   2158 INPUT
- A1: ART COMPLEX TECH COMPLEX
- A2: ARCHAIC SIMPLE ARCHAIC SIMPLE---BEGINNING OF A NEW COPERNICAN AGE
- B1: PHYS TECH TECH CREATIVE
- B2: MATH MATH BASIC MATH
- C1: TECH PHYS TECH PHYS NEW COPERNICAN AGE
- C2: ABSTRACT SUBJECTIVE SIMPLE ABSTRACT

YEAR =   2159 INPUT
- A1: ART COMPLEX TECH SIMPLE
- A2: ARCHAIC COMPLEX ARCHAIC SIMPLE
- B1: PHYS TECH TECH TECH
- B2: MATH MATH BASIC BASIC
- C1: TECH UNIVERSE TECH PHYS
- C2: ABSTRACT SUBJECTIVE SIMPLE SUBJECTIVE

YEAR =   2160 INPUT
- A1: ART COMPLEX ART SIMPLE
- A2: MATH COMPLEX ARCHAIC SIMPLE
- B1: PHYS TECH CREATIVE TECH
- B2: MATH MATH MATH BASIC
- C1: CREATIVE UNIVERSE TECH PHYS
- C2: ABSTRACT SUBJECTIVE MATH SUBJECTIVE

YEAR =   2161 INPUT
- A1: ART COMPLEX ART COMPLEX
- A2: MATH SIMPLE ARCHAIC SIMPLE
- B1: PHYS TECH CREATIVE CREATIVE
- B2: MATH MATH MATH MATH
- C1: CREATIVE PHYS TECH PHYS
- C2: ABSTRACT SUBJECTIVE MATH ABSTRACT

YEAR =   2162 INPUT
- A1: COSM COMPLEX ART COMPLEX
- A2: SUBJECTIVE SIMPLE ARCHAIC SIMPLE
- B1: PHYS TECH UNIVERSAL CREATIVE
- B2: SPECIAL MATH MATH MATH
- C1: UNIVERSAL PHYS TECH PHYS
- C2: ABSTRACT SUBJECTIVE SUBJECTIVE ABSTRACT

YEAR =   2163 INPUT

A1: **COSM SIMPLE ART COMPLEX**
A2: **SUBJECTIVE COMPLEX ARCHAIC SIMPLE**
B1: **PHYS TECH UNIVERSAL TECH**
B2: **SPECIAL BASIC MATH MATH**
C1: **UNIVERSAL UNIVERSE TECH PHYS**
C2: **ABSTRACT SUBJECTIVE SUBJECTIVE SUBJECTIVE**

YEAR =   2164 INPUT

A1: **PHYS SIMPLE ART COMPLEX**
A2: **ABSTRACT COMPLEX ARCHAIC SIMPLE**
B1: **PHYS TECH PHYS TECH**
B2: **ABSTRACT BASIC MATH MATH**
C1: **PHYS UNIVERSE TECH PHYS**
C2: **ABSTRACT SUBJECTIVE ABSTRACT SUBJECTIVE**

YEAR =   2165 INPUT

A1: **PHYS COMPLEX ART COMPLEX**
A2: **ABSTRACT SIMPLE ARCHAIC SIMPLE**
B1: **PHYS CREATIVE PHYS TECH**
B2: **ABSTRACT MATH MATH MATH**
C1: **PHYS PHYS TECH PHYS**
C2: **ABSTRACT ABSTRACT ABSTRACT SUBJECTIVE**

YEAR =   2166 INPUT
    A1: NEW INVENTION ART COMPLEX
    A2: GROUP CONCEPT ARCHAIC SIMPLE
    B1: INTELLIGENT COMPLEXITY PHYS TECH
    B2: STUPIDLY SIMPLE MATH MATH
    C1: INTELLIGENT CREATIVE TECH PHYS
    C2: BASIC TECHNIQUE ABSTRACT SUBJECTIVE

YEAR =   2167 INPUT
    A1: COSM COMPLEX TECH COMPLEX
    A2: ARCHAIC COMPLEX ARCHAIC SIMPLE
    B1: PHYS CREATIVE TECH CREATIVE
    B2: SPECIAL MATH BASIC MATH
    C1: TECH UNIVERSE TECH PHYS
    C2: ABSTRACT ABSTRACT SIMPLE ABSTRACT

YEAR =   2168 INPUT
    A1: COSM COMPLEX TECH SIMPLE
    A2: ARCHAIC COMPLEX ARCHAIC COMPLEX
    B1: PHYS CREATIVE TECH TECH
    B2: SPECIAL MATH BASIC BASIC
    C1: TECH UNIVERSE TECH UNIVERSE
    C2: ABSTRACT ABSTRACT SIMPLE SUBJECTIVE

YEAR =   2169 INPUT
   A1: COSM COMPLEX ART SIMPLE
   A2: MATH COMPLEX ARCHAIC COMPLEX
   B1: PHYS CREATIVE CREATIVE TECH
   B2: SPECIAL MATH MATH BASIC
   C1: CREATIVE UNIVERSE TECH UNIVERSE
   C2: ABSTRACT ABSTRACT MATH SUBJECTIVE

YEAR =   2170 INPUT
   A1: COSM COMPLEX ART COMPLEX
   A2: MATH COMPLEX ARCHAIC SIMPLE
   B1: PHYS CREATIVE CREATIVE CREATIVE
   B2: SPECIAL MATH MATH MATH
   C1: CREATIVE UNIVERSE TECH PHYS
   C2: ABSTRACT ABSTRACT MATH ABSTRACT

YEAR =   2171 INPUT
   A1: COSM COMPLEX COSM COMPLEX
   A2: SUBJECTIVE SIMPLE ARCHAIC COMPLEX
   B1: PHYS CREATIVE UNIVERSAL CREATIVE
   B2: SPECIAL MATH SPECIAL MATH
   C1: UNIVERSAL PHYS TECH UNIVERSE
   C2: ABSTRACT ABSTRACT SUBJECTIVE ABSTRACT

YEAR =   2172 INPUT
    A1: **COSM SIMPLE COSM COMPLEX**
    A2: **SUBJECTIVE COMPLEX ARCHAIC COMPLEX**
    B1: **PHYS CREATIVE UNIVERSAL TECH**
    B2: **SPECIAL BASIC SPECIAL MATH**
    C1: **UNIVERSAL UNIVERSE TECH UNIVERSE**
    C2: **ABSTRACT ABSTRACT SUBJECTIVE SUBJECTIVE**

YEAR =   2173 INPUT
    A1: **PHYS SIMPLE COSM COMPLEX**
    A2: **ABSTRACT COMPLEX ARCHAIC COMPLEX**
    B1: **PHYS CREATIVE PHYS TECH**
    B2: **ABSTRACT BASIC SPECIAL MATH**
    C1: **PHYS UNIVERSE TECH UNIVERSE**
    C2: **ABSTRACT ABSTRACT ABSTRACT SUBJECTIVE**

YEAR =   2174 INPUT
    A1: **PHYS COMPLEX COSM COMPLEX**
    A2: **ABSTRACT SIMPLE ARCHAIC COMPLEX**
    B1: **PHYS CREATIVE PHYS CREATIVE**
    B2: **ABSTRACT MATH SPECIAL MATH**
    C1: **PHYS PHYS TECH UNIVERSE**
    C2: **ABSTRACT ABSTRACT ABSTRACT ABSTRACT**

YEAR =   2175 INPUT

    A1: NEW INVENTION COSM COMPLEX

    A2: GROUP CONCEPT ARCHAIC COMPLEX

    B1: INTELLIGENT COMPLEXITY PHYS CREATIVE

    B2: STUPIDLY SIMPLE SPECIAL MATH

    C1: INTELLIGENT CREATIVE TECH UNIVERSE

    C2: BASIC TECHNIQUE ABSTRACT ABSTRACT

YEAR =   2176 INPUT

    A1: COSM SIMPLE TECH COMPLEX

    A2: MATH COMPLEX ARCHAIC SIMPLE

    B1: INTELLIGENT COMPLEXITY TECH CREATIVE

    B2: SPECIAL BASIC BASIC MATH

    C1: CREATIVE UNIVERSE TECH PHYS

    C2: BASIC TECHNIQUE SIMPLE ABSTRACT

YEAR =   2177 INPUT

    A1: COSM SIMPLE TECH SIMPLE

    A2: MATH COMPLEX ARCHAIC COMPLEX

    B1: INTELLIGENT COMPLEXITY TECH TECH

    B2: SPECIAL BASIC BASIC BASIC

    C1: CREATIVE UNIVERSE TECH UNIVERSE

    C2: BASIC TECHNIQUE SIMPLE SUBJECTIVE

YEAR =   2178 INPUT
   A1: COSM SIMPLE ART SIMPLE
   A2: MATH COMPLEX MATH COMPLEX
   B1: INTELLIGENT COMPLEXITY CREATIVE TECH
   B2: SPECIAL BASIC MATH BASIC
   C1: CREATIVE UNIVERSE CREATIVE UNIVERSE
   C2: BASIC TECHNIQUE MATH SUBJECTIVE

YEAR =   2179 INPUT
   A1: COSM SIMPLE ART COMPLEX
   A2: MATH SIMPLE MATH COMPLEX
   B1: INTELLIGENT COMPLEXITY CREATIVE CREATIVE
   B2: SPECIAL BASIC MATH MATH
   C1: CREATIVE PHYS CREATIVE UNIVERSE
   C2: BASIC TECHNIQUE MATH ABSTRACT

YEAR =   2180 INPUT
   A1: COSM SIMPLE COSM COMPLEX
   A2: SUBJECTIVE SIMPLE MATH COMPLEX
   B1: INTELLIGENT COMPLEXITY UNIVERSAL CREATIVE
   B2: SPECIAL BASIC SPECIAL MATH
   C1: UNIVERSAL PHYS CREATIVE UNIVERSE
   C2: BASIC TECHNIQUE SUBJECTIVE ABSTRACT

YEAR =   2181 INPUT
    A1: COSM SIMPLE COSM SIMPLE
    A2: SUBJECTIVE COMPLEX MATH COMPLEX
    B1: INTELLIGENT COMPLEXITY UNIVERSAL TECH
    B2: SPECIAL BASIC SPECIAL BASIC
    C1: UNIVERSAL UNIVERSE CREATIVE UNIVERSE
    C2: BASIC TECHNIQUE SUBJECTIVE SUBJECTIVE

YEAR =   2182 INPUT
    A1: PHYS SIMPLE COSM SIMPLE
    A2: ABSTRACT COMPLEX MATH COMPLEX
    B1: INTELLIGENT COMPLEXITY PHYS TECH
    B2: ABSTRACT BASIC SPECIAL BASIC
    C1: PHYS UNIVERSE CREATIVE UNIVERSE
    C2: BASIC TECHNIQUE ABSTRACT SUBJECTIVE

YEAR =   2183 INPUT
    A1: PHYS COMPLEX COSM SIMPLE
    A2: ABSTRACT SIMPLE MATH COMPLEX
    B1: INTELLIGENT COMPLEXITY PHYS CREATIVE
    B2: ABSTRACT MATH SPECIAL BASIC
    C1: PHYS PHYS CREATIVE UNIVERSE
    C2: BASIC TECHNIQUE ABSTRACT ABSTRACT

TECH PROPHECIES 2100 TO 2199 / NATHAN COPPEDGE

YEAR =   2184 INPUT
- A1: NEW INVENTION COSM SIMPLE
- A2: GROUP CONCEPT MATH COMPLEX
- B1: INTELLIGENT COMPLEXITY INTELLIGENT COMPLEXITY END OF A COPERNICAN AGE
- B2: STUPIDLY SIMPLE SPECIAL BASIC
- C1: INTELLIGENT CREATIVE CREATIVE UNIVERSE
- C2: BASIC TECHNIQUE BASIC TECHNIQUE END OF A COPERNICAN AGE

YEAR =   2185 INPUT
- A1: PHYS SIMPLE TECH COMPLEX
- A2: MATH SIMPLE ARCHAIC SIMPLE
- B1: TECH CREATIVE TECH CREATIVE BEGINNING OF A NEW COPERNICAN AGE
- B2: ABSTRACT BASIC BASIC MATH
- C1: CREATIVE PHYS TECH PHYS
- C2: SIMPLE ABSTRACT SIMPLE ABSTRACT BEGINNING OF A NEW COPERNICAN AGE

YEAR =   2186 INPUT
- A1: PHYS SIMPLE TECH SIMPLE
- A2: MATH SIMPLE ARCHAIC COMPLEX
- B1: TECH TECH TECH CREATIVE
- B2: ABSTRACT BASIC BASIC BASIC
- C1: CREATIVE PHYS TECH UNIVERSE
- C2: SIMPLE SUBJECTIVE SIMPLE ABSTRACT

YEAR =   2187 INPUT
    A1: PHYS SIMPLE ART SIMPLE
    A2: MATH SIMPLE MATH COMPLEX
    B1: CREATIVE TECH TECH CREATIVE
    B2: ABSTRACT BASIC MATH BASIC
    C1: CREATIVE PHYS CREATIVE UNIVERSE
    C2: MATH SUBJECTIVE SIMPLE ABSTRACT

YEAR =   2188 INPUT
    A1: PHYS SIMPLE ART COMPLEX
    A2: MATH SIMPLE MATH SIMPLE
    B1: CREATIVE CREATIVE TECH CREATIVE
    B2: ABSTRACT BASIC MATH MATH
    C1: CREATIVE PHYS CREATIVE PHYS
    C2: MATH ABSTRACT SIMPLE ABSTRACT

YEAR =   2189 INPUT
    A1: PHYS SIMPLE COSM COMPLEX
    A2: SUBJECTIVE SIMPLE MATH SIMPLE
    B1: UNIVERSAL CREATIVE TECH CREATIVE
    B2: ABSTRACT BASIC SPECIAL MATH
    C1: UNIVERSAL PHYS CREATIVE PHYS
    C2: SUBJECTIVE ABSTRACT SIMPLE ABSTRACT

YEAR =   2190 INPUT
    A1: PHYS SIMPLE COSM SIMPLE
    A2: SUBJECTIVE COMPLEX MATH SIMPLE
    B1: UNIVERSAL TECH TECH CREATIVE
    B2: ABSTRACT BASIC SPECIAL BASIC
    C1: UNIVERSAL UNIVERSE CREATIVE PHYS
    C2: SUBJECTIVE SUBJECTIVE SIMPLE ABSTRACT

YEAR =   2191 INPUT
    A1: PHYS SIMPLE PHYS SIMPLE
    A2: ABSTRACT COMPLEX MATH SIMPLE
    B1: PHYS TECH TECH CREATIVE
    B2: ABSTRACT BASIC ABSTRACT BASIC
    C1: PHYS UNIVERSE CREATIVE PHYS
    C2: ABSTRACT SUBJECTIVE SIMPLE ABSTRACT

YEAR =   2192 INPUT
    A1: PHYS COMPLEX PHYS SIMPLE
    A2: ABSTRACT SIMPLE MATH SIMPLE
    B1: PHYS CREATIVE TECH CREATIVE
    B2: ABSTRACT MATH ABSTRACT BASIC
    C1: PHYS PHYS CREATIVE PHYS
    C2: ABSTRACT ABSTRACT SIMPLE ABSTRACT

YEAR =   2193 INPUT

- A1: NEW INVENTION PHYS SIMPLE
- A2: GROUP CONCEPT MATH SIMPLE
- B1: INTELLIGENT COMPLEXITY TECH CREATIVE
- B2: STUPIDLY SIMPLE ABSTRACT BASIC
- C1: INTELLIGENT CREATIVE CREATIVE PHYS
- C2: BASIC TECHNIQUE SIMPLE ABSTRACT

YEAR =   2194 INPUT

- A1: PHYS COMPLEX TECH COMPLEX
- A2: SUBJECTIVE SIMPLE ARCHAIC SIMPLE
- B1: TECH TECH TECH CREATIVE
- B2: ABSTRACT MATH BASIC MATH
- C1: UNIVERSAL PHYS TECH PHYS
- C2: SIMPLE SUBJECTIVE SIMPLE ABSTRACT

YEAR =   2195 INPUT

- A1: PHYS COMPLEX TECH SIMPLE
- A2: SUBJECTIVE SIMPLE ARCHAIC COMPLEX
- B1: TECH TECH TECH TECH
- B2: ABSTRACT MATH BASIC BASIC
- C1: UNIVERSAL PHYS TECH UNIVERSE
- C2: SIMPLE SUBJECTIVE SIMPLE SUBJECTIVE

YEAR =   2196 INPUT
    A1: PHYS COMPLEX ART SIMPLE
    A2: SUBJECTIVE SIMPLE MATH COMPLEX
    B1: CREATIVE TECH TECH TECH
    B2: ABSTRACT MATH MATH BASIC
    C1: UNIVERSAL PHYS CREATIVE UNIVERSE
    C2: MATH SUBJECTIVE SIMPLE SUBJECTIVE

YEAR =   2197 INPUT
    A1: PHYS COMPLEX ART COMPLEX
    A2: SUBJECTIVE SIMPLE MATH SIMPLE
    B1: CREATIVE TECH TECH CREATIVE
    B2: ABSTRACT MATH MATH MATH
    C1: UNIVERSAL PHYS CREATIVE PHYS
    C2: MATH SUBJECTIVE SIMPLE ABSTRACT

YEAR =   2198 INPUT
    A1: PHYS COMPLEX COSM COMPLEX
    A2: SUBJECTIVE SIMPLE SUBJECTIVE SIMPLE
    B1: UNIVERSAL CREATIVE TECH TECH
    B2: ABSTRACT MATH SPECIAL MATH
    C1: UNIVERSAL PHYS UNIVERSAL PHYS
    C2: SUBJECTIVE ABSTRACT SIMPLE SUBJECTIVE

YEAR =   2199 INPUT
    A1: **PHYS COMPLEX COSM SIMPLE**
    A2: **SUBJECTIVE COMPLEX SUBJECTIVE SIMPLE**
    B1: **UNIVERSAL TECH TECH TECH**
    B2: **ABSTRACT MATH SPECIAL BASIC**
    C1: **UNIVERSAL UNIVERSE UNIVERSAL PHYS**
    C2: **SUBJECTIVE SUBJECTIVE SIMPLE SUBJECTIVE**

RECOMMENDED READING

The Black Swan Market

The History of Perpetual Motion Machines

The History of Coherence

Bio

Nathan Larkin Coppedge is a
philosopher, artist, inventor, and
poet who lives in New Haven,
Connecticut near Yale.

www.ingramcontent.com/pod-product-compliance
Lightning Source LLC
Chambersburg PA
CBHW060011300526
45794CB00003B/1169